Body Fodder

By

Ron Hine

Copyright © 2024 The HAHA Train

All rights reserved.

ISBN:

DEDICATION

As shared by scholars and other leaders, the more you share a physical object, the more fragmented, used, and weaker it becomes. Inversely, the more you share an idea, the stronger and more resilient it becomes. This book is dedicated to the mind because it will always rule over the body, no matter how much the body complains.

ACKNOWLEDGMENTS

What did we do before the internet? To all the wonderful sources of information about the body that are out there, thank you! Here are a few of them: National Library of Medicine, National Library of Science, Ask Dr. Universe, Hartford Healthcare, Select Health, INTEGRIS Health, Pfizer, Northwestern Medicine, Science Learning Hub, Listverse, Wikipedia, WebMD, and VCA Hospitals. So many sources, so little time.

INTRO TO BODY FODDER

The more one examines the physical world, the weirder it becomes. Forget about our political fireworks, trending TikTok videos, or that interesting couple with the elaborate shrunken head collection that just moved in next door (and runs a daycare). You don't have to look any farther than the mirror to see the most amazing and strangest show on earth. Whether your buff or a jolly bowl of jelly, the basic geometry, mechanics, and universal influences on the body are about the same for everyone. The geometry of the body seems to follow some organic rules of alignment and uniqueness, as if the great designer was guided by preprinted instructions (but didn't fill out the warranty card). While we sip our morning latte, attempt our favorite yoga pose, or stress out over that report we forgot to send our boss, our body machine is performing over a sextillion (a billion trillion) chemical reactions every second. And that doesn't include the billions of other functions the body is busy executing without any help from us (and it never asks for a tip). The body is an amazing universe that is reacting to our thoughts and interacting with the world around us at a level we never see (I know, more spying we need to consider). This book is a collection of some human body facts, so you are armed for that next family and friend's barbeque. If nothing else, this book may finally give you the information you need to convince your friends that, while your gas problem may have disturbed their dinner party, your flatus expressions are a completely natural part of being human.

THE POWER OF WE

It's a known fact (by all atoms) that each time you exhale, you breathe out ten to the power of twenty-two atoms of your internal organs. When you inhale, you breathe in ten to the power of twenty-two atoms of other people. We are breathing in and out each other. It seems easier to accept the vague and elusive idea that we might be one in consciousness. Considering all the unpleasant bodily functions and fluids we must accept as human beings, I'm not sure we are ready to accept we are swapping our guts with each other. The subsequent emotion can be anything from elated that we're partially made of Taylor Swift, to terrified that Gary Ridgeway is lurking inside us. We not only incorporate other people's atoms from breathing but also from eating and drinking. This exchange with our fellow humans is a little deeper than just sipping our friend's invisible slime. Some estimates declare that we have hundreds of billions of atoms that, only a year ago, were inside every other person on earth. So, who needs to travel! A part of you has already been everywhere. Discovering this crazy fact can be frustrating. It seems that science is now backing up my mother's advice that when I hate someone, I'm only hating myself.

LESSER STRESSOR

Some studies show 75-90% of illnesses and diseases are either caused or complicated by stress. Wow doesn't that complicate life. If your job is your biggest stress and you quit, you'll stress about money and being broke. If people stress you out, isolation can cause stress. A dog might help some, but they're not great conversationalist, for most people. Maybe your stressors are rollercoasters, babies, politicians, or clowns (which are basically all the same thing). Eventually, you will be confronted with them. Perhaps stress is the real reason we grow old and exit this world. While we are young and distracted, we don't have time to dwell on our stress. When later in life we have time to think about stress, maybe we just determine that life isn't worth picking up other people's dog poop or, letting those neighborhood kids run across our new lawn (although, seeing those kids step in the dog poop may bring some stress relief). No doubt the corporate powers of the world don't want you to believe that stress has such a dire impact on health. The litigation possibilities are enough to make an injury lawyer giddy. However, don't set your court date yet. Since one person's catastrophe can be another person's best day ever, the source of stress will remain in the mind of the frazzled.

BODY GEOMETRY

Like the rest of the universe, the human body has some interesting rules that it must follow. Some will say these rules come from a creator. Some will say evolution is to blame. No doubt someone will come up with a conspiracy theory like we are assembled from parts by aliens, like a Cootie (please don't post that on social media). Here is an interesting rule. With few exceptions, a human's nipples are the same distance apart as their earlobes. No one seems to know why. However, this does sounds like one of those great studies that some government entity will eventually pour money into. Here is another one. You are as tall as your arm span. Why is that?! Is it a creator trying to keep things fair by keeping them even? Or is it mother nature ensuring that the picture is never hanging crooked? Alas, the "why" question can never be fully answered. What really matters is that you check your earlobes, nipples, arm length, and height right now to confirm this strange human geometry rule. I know you want to.

BLOODY COMMUTE

If you think your commute is bad, try being a blood cell. These little oxygenators travel about 12,000 miles per day to deliver their little O2 packages all over your body, free of charge (no membership required). Since most of us are no more than five to six feet tall, 12,000 miles is a dizzying amount of laps from your heart, around your big toe, and back to your pumper again. These little red hemoglobin delivery agents don't complain or ask for days off. They happily visit every muscle, organ, nerve, and flimsy piece of tissue to deliver their precious cargo. They not only deliver your life-giving oxygen. They've also cornered the food delivery circuit, showing up to every cell door with tasty nutrients, no phone number or mobile app needed. Not only that, but they handle all the package returns as well, picking up carbon dioxide and dropping it off at the lungs. Just in time for the next flight out on exhale air. This highly efficient and precise delivery and returns processing engine has achieved a level of logistics perfection that would make even Amazon jealous. Let's just hope they don't try to unionize. I'm not sure what they might ask for in the negotiations but, it's clear we'd never survive a strike.

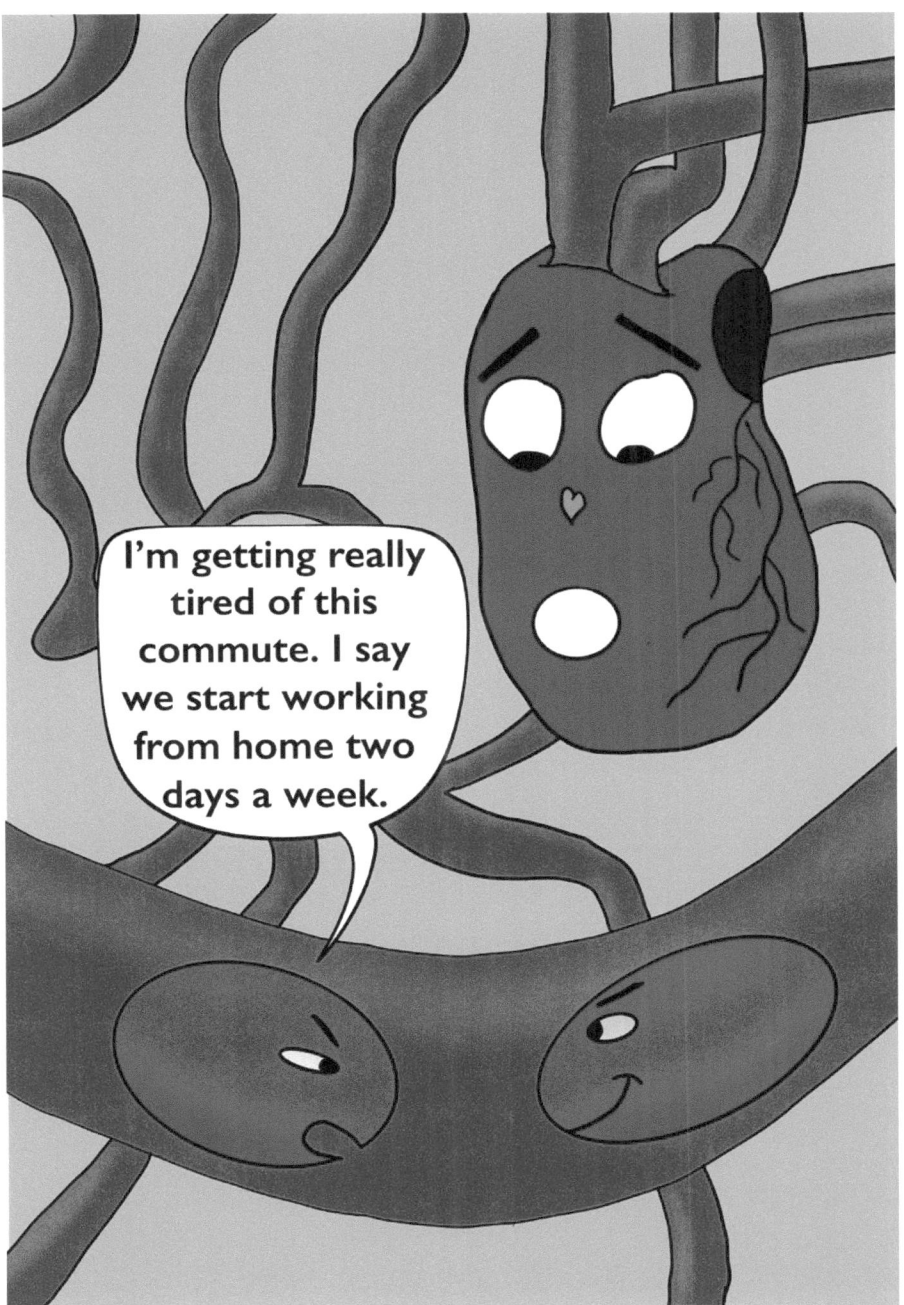

A STOMACH OF STEEL

Oh, the things we humans put into our stomachs. Everything from, heavenly sweet strawberries and ice cream (popular in grandma's kitchen), to the large fried spiders served as a crunchy entre in some Asian cultures. Yes, you name it, and some human somewhere will eat it. When you think of a human stomach, you might think of soft pink and white tissue that gently receives your chosen delicacies, peacefully disassembles them to harvest the nutrients, and politely sends the waste along its way. No way! It's a war down there! It's literally the battle of the bacteria. But the fiercest warrior in the cave, the meanest menace ready to attack all who enter, is stomach acid. Hydrochloric acid to be exact. Acidity is measured as a PH of 1 to 14 with one being the highest. Our little digestor produces about 1.5 liters a day of an acid soup that's a whopping 1.5 to 3.5 PH. This PH is high enough to dissolve metal! This seemed like useless information until I read about Monsieur Mangetout in the Guinness Book of World Records. Monsieur ate about two pounds of metal each day (which is much higher than the recommended daily dose of iron). I imagine Mr. Mangetout went through a lot of toilets. In any case, the next time you eat something and you're a little worried that your stomach can't handle it, f-o-r-g-e-t-a-b-o-u-t-i-t. Your stomach has an arsenal ready to destroy any dietary enemy you happen to engage at your favorite auto parts buffet.

A TASTIER IDENTITY

With the continuous onslaught of technology innovations that rain down on our existence, humans have developed plenty of ways to validate a person's identity. Meeting face to face is no longer necessary. Besides, viruses have taken the fun out of that technique. The days of looking at a person's face and comparing it with an ID are quickly being kicked to the curb by digital fingerprints, retina scans, facial recognition, and anus scans (ok, I made up the last one, but someone should research it). I'm guessing you could slap about any internal organ on a scanner and get a unique pattern. However, that would be messy and be even more invasive than social media marketing, if that's possible. One very convenient method of identification that has yet to catch on is tongue prints. Yes, everyone has a unique tongue pattern! This is not surprising considering the varying degrees of tastes in our societies. In one sense, I'm surprised tongues have not become a more prevalent identification option. When you consider all the growing conflict in the world, it seems like people would be more than willing to stick their tongue out at the authorities. Maybe someday, our latent lickers will be called upon to verify our unique cellular makeup. For now, most technologies that involve spit to operate them are not likely to catch on, unless sex is involved.

MONDAY MORNING WARNING

Around and around we go. Work, eat, sleep, repeat. Statistically speaking, we are stressed at work. If you are one of the few fortunate that really love your job, you may live to a ripe old age (note: you cannot fool your body). For most of us, our work is like a newly released album that has one great song, and the rest is crap. We trudge through the sludge every day waiting for our boss to throw us a fun project bone. We are the only species that, every day, just keeps migrating to a place we'd like to avoid (most animal species can learn not to do that). Studies have shown that working people have the highest risk of a heart attack on Monday morning. You would think that this fact would give more momentum to the idea of a four-day work and push Monday into the weekend bucket. Think of how many lives could be saved! As we have all learned from governments and big business, saving lives is very very important, unless it cost too much money. All trusty HR departments will tell you that having fun at work, so you can live longer and be healthier, is their top priority. Your spidey sense will tell you not to let fun get in the way of that new boat your boss is planning to buy with their bonus. So, the next time you want to tack on an extra day off for a long weekend, forget about Friday. If not for yourself then for the sake of your loved ones, take Monday off.

WHAT'S YOUR STINK

There is a reason that dogs smell butts, and it's not just because they are conveniently located at a good height. Smelling each other is just their way of checking doggie identification. Some experts say that dogs can tell age, weight, gender, temperament, diet, and movie preferences with just a quick rear orifice inspection. Ok, maybe they didn't say movie preferences but, why not?! In fact, I think it goes further than that because humans have a completely unique stink! That's right! The exception to the rule is twins. They smell the same (which is probably confusing for dogs). I image that back in the days of the Neanderthal (before shaking hands became a thing), sniffing someone you just met was a perfectly acceptable way to greet them. In fact, since showers were in relatively short supply for cave dwellers, you probably knew who was coming to dinner when they were within a hundred yards. I suppose once bathing started to catch on, this handy form of identification fell apart. No longer could you count on checking a person's stink ID. You had to figure it out by sight, which was difficult because your hair was so long, you could hardly see anyone. And even if you could, everyone just looked like a big hairball. It's easy to see why we eventually needed little plastic ID cards (and barbers). So, the next time you are pulled over by the police and they ask for your ID, just tell them they can get a lot more information by smelling your butt instead. Then call a lawyer.

EXPRESS THY FINGER

Oh, that middle finger is such an expressive little rebel. No other finger on the hand can express states of outrage, or significantly increase your risk of getting shot, quite like the middle meddler. According to my research on the middle finger (no telling how that research will affect the marketing adds I get), you can blame the ancient Greeks for casting this dark shadow on your middle digit. The gesture's phallic meaning is not a casual coincidence. Some say the symbol is intended to represent the complete male genital. Really? My next question is, who's? It appears that the elevated importance bestowed on this celebrity appendage is not lost on our biology. The fastest growing fingernail on the hand is, you guessed it, the middle finger. Forget the fact that you lose 50% of your hand strength without your pinky finger. That little runt can't get any respect compared to its boisterous middle sibling. That middle finger will just be quietly hanging out. Then, like a rock star stepping into a crowd of paparazzi, as the magnificent middle raises its shinny nail to the heavens, you can hear the gasps, anger, fainting spells, and laughter from just about anyone with a good visual. So, the next time that person in the car beside you flips you the bird for cutting them off, don't get mad. Remember they are sharing with you a prestigious symbol that has survived the test of time and carries a rich heritage. Then you can flip them the bird, with a smile.

SUPER SCHNOZZLE TO THE RESCUE

If you ask someone what's the nose for, the optimist with likely reflect upon the lip-smacking aroma of that casserole that granny used to make, or that fine wine they love with just a hint of cherry notes (I can never taste the notes). The pessimist will likely say it is for avoiding dangerous smells, like the crap the neighbor's dog leaves on your lawn. Yes, we typically think of that protruding cartilage cave as our tool for sampling the fragrances of the world, good or bad. People don't typically view their nose for what it truly is, a superhero! Yes, it's a bone, it's a cartilage, it's Super Snotty! This facial extension only looks like a harmless place to hang glasses. In reality, it is protecting you from the villains of the particle cartel. Yes, those mules of dirt, dust, and bacteria are constantly attempting to cross the body boarder and deposit their filthy packages in your membranes for distribution to your cellular communities. That's when the nostril crusader springs into action. Your super snout quickly assembles a task force of abdominal muscles, lungs, and tongue. Using hurricane strength one hundred mile per hour winds, your hooter boots those criminals from your body (accompanied by superhero sounds like snort, sniff, and a-chew). So remember, you can believe smelling is the most important function of the nose, but it's snot.

Body Fodder

HEART GOT THE BEAT

There is a good reason that you want to listen to fast music when executing your favorite fat to muscle conversion routine or, listen to slow music when sitting by the pond sipping chardonnay and watching the leaves rustle. The music not only sets the mood for the adventure du jour. It acts as a foot on the gas pedal of your metabolism by speeding up or slowing down your heartbeat. That's right, your heart will beat to mimic the beat of the music funneling through your EarPods. Of course, since our body functions always want to party together, your respiratory system also speeds up or slows down with your heart rate. And like a groupie in love with the bass player, your blood pressure will also dance fast or slow with the party crowd. Because this is a physiological response, it requires the observer perceive the beat as fast "music". For most creatures, fast music that pushes your body into a state that resembles anxiety is just harmful noise and cause to run away. But not humans. Us homo sapiens are experts at transforming anxiety into an exciting experience (and then charging admission). This is what makes us an advanced species, maybe. Just make sure you don't fall asleep listening to a-ha's Take on Me and accidentally hold down the fast forward. You may wake up in the sweaty pool of an alternate dimension. Or at least a sweaty pool.

THE HAIRY TRUTH

Either secretly or obnoxiously obvious, many of us humans are obsessed with how we look. We toil for hours at the gym, or endlessly search for the perfect outfit that doesn't make us look too fat, or too skinny, but just right. But after it's put together as good as it's going to get below the neck, we turn our attention to the crowning glory, our hair. Some of us are topped with beautiful wavy locks that hang with us until the party is over. However, for many men, that playfully sadistic magician of nature, named gene, waves his diabolical wand and our hair begins to disappear. Great, now we look like a well-dressed que ball. Initially, we see a well-stocked hat collection as our only refuge. Unfortunately, the hat must eventually come off at the worst times (fancy restaurants, church, fiancé's parent's dinner table, etc.). But don't fret my shinny capped brethren. For the mop on top is not the king of the furry forest. The fastest growing hair on the human body is that cousin on the noggin's lower 40, the beard. That's right! Nothing offsets that glowing topline deficit like a gorgeous chin jacket. And there are so many styles to choose from! Beards can be as noble as Abe Lincoln, as manly as Paul Bunyan, or as sexy as Chris Hemsworth. Ok, I know what you're saying, all those guys also have hair on top. That's beside the point. Nothing says "oh, I do too have hair" like a well-groomed fur lined lips waiting to securely catch that next kiss.

Body Fodder

FAST THINKING

Thinking is just something we all do (I think). We take our thought sequences for granted as we follow them down their logic pathways, deciding which are great ideas and which are not so great. Once we decide that an idea is worthy, we convert it into action. While the eventual physical step we take is comparatively slow (especially in the morning before coffee), the thinker process going on in the brain is a regular Sonic the Hedge Hog. That crazy idea you just had (like, let's see how fast this car goes!) shows up as light on your neuropathway (and a glimmer in your eye). Once you've lit up that idea, that little thought beam steps on the gas pedal and races down that neuro highway at a blinding 250 miles per hour (luckily, there are no neuronal police). And just like every good interstate highway, this cerebral transportation infrastructure is always under construction. Every time you learn something new, your brainy little workers break out their equipment and build another roadway, just in time for those approaching head lights to blast past like a formula one racecar (there are no slow construction zones on this knowledge highway). And if you decide you like that idea and revisit it regularly, those little racers will just keep going down that same track, no tire changes required. So remember, you don't need to wait for congress to pass any new infrastructure bills. Just read something new and you'll bask in the intellectual wealth of a thriving neuronal construction business.

Body Fodder

ALIEN BODY INVASION

Every time I start feeling a little human again, along comes another piece of information that convinces me otherwise. For example, the average human body is 60% water (I knew my fat was mostly water weight). If you feel like a big nothing at times, it's because your body is 99.99% empty space! This is because most of an atom's volume is space (note: atom space is really a bunch of quantum stuff that's tough to understand so, it might as well be space). It's easy to accept that our body includes friendly comrades like water and space. What's unnerving is that our body has about the same number of bacteria and microbes as it does human cells. I say what?! This lovely fact can really make you feel half human (without drinking any alcohol). OK, so maybe we should accept the fact that this body is only half ours. But I can't! I know our cells have been living in harmony with these strange bacterial aliens for thousands of years but, so what! Eventually all great civilizations come to an end, right? All it took was social media and a little anger to split the United States like two butt cheeks getting the worst wedgie in history. Remember your last infection? We all know how angry bacteria can get and I'm sure they figured out the social media piece by now (I think they might call it bac-talk). All we can do for now is stockpile B vitamin ammunition for the immune system revolution. I can only assume that bacteria and human cell negotiations in my body are going ok for now. If talks break down, I'm sure I'll be the first to know.

TAKE IT LYING DOWN

As humans age, most begin to shrink. Yes, the speed varies (depending on how well your parents pushed good posture). But for those getting to the point where gravity has begun to win the tug of war, it can be quite a shock when you first get the news. You're usually at the doctor's office when they tell you you're a quarter inch shorter. At first, you'll blame it on the thicker socks you wore to the last appointment. As it moves to a half inch, then three quarters, you'll argue with the nurse that their measuring device is defective. As you approach the shrinkage of an inch, you'll panic and begin your own measurements at home. Once you've lost more than an inch of your hard-earned height, you'll begin to accept your slow descent to the earth and thick soled shoes become your new BFF. But, if a small thing like the loss of a quarter inch can send you into gravitational despair, fear not. For every night, as you lay in bed, unlike the bank accounts of the greedy, the body beautiful gives back a little to humanity. About one centimeter to be exact (a little less than a half inch). As you lay in bed staring at the celling worrying about how small you'll be tomorrow, your cartilage is stretching back out to its original form. What fantastic news! This makes every morning a big day! At least, a little bigger than yesterday evening. Alas, by evening you'll have shrunk back to your small self. So, seniors listen up. Make your doctor appointments in the morning. That way, life will seem just a little larger.

WHO ARE WE ANYWAY?

Everything in this crazy body is in a constant state of transition. If you don't like change, fasten your biological seat belt, because you're in for a carnival ride that might make you puke. Sure, things like rotating fashion styles, trending food fads, and cell phone app updates are the obvious change agents that we either diligently embrace or blissfully ignore. We also must deal with the fickleness of our own mind exploration, as our favorite likes ( ) of today slowly transition to our rant targets of tomorrow (and versa visa). But nothing is as freaky as knowing the body I looked at in the mirror a year ago is different than the body I see today. 98% different to be exact. Yes, science estimates that 98% of the atoms in your body are replaced with new atoms each year. For many, this fact typically falls in the ignorantly ignore category. So, what can you do with this information? If you tell the IRS that someone else earned that money last year, your tax bill will likely be much higher from the fines after the audit. If this "not me" argument ever makes it to the Supreme Court, we're sure to have a constitutional crisis. Taxes aside, one of the biggest questions I have is, why are we getting older?! If a new set of atoms arrive each year, don't we get a do over? When these replacement atoms show up to check in, who's giving them the same old body to wear? Shouldn't they bring their own luggage? I find this very suspicious. I think someone is talking my new atoms into putting on my old wrinkles. I hope it isn't me.

SEEING IS A POWERFUL THING

Can you see me now? Of course you can. If the lid is open, the human eye is collecting all the visible world's data for further examination. The eye captures light and converts it to impulses that are fed to the brain for interpretation (where the brain cells ask each other, can we believe those eyes?). Setting aside our brains for the moment, the eyes are an amazing feat. Eyes can see a teeny tiny single photon, which is the smallest unit of light. Don't ask how big a photon is, you won't like the answer (unless your good at physics). You can't measure a photon because it has no mass (see, I told you). And if that makes your eyes exhausted, they aren't. Eyes function at 100% of their ability when the shutter is open and never get fatigued. If you think your eyes are tired, check in with your mind. It's probably the one who needs a nap. Your eyes can process and send 10 bits of information to the brain every second, 600 bits every minute, and 36,000 bits every hour. With that kind of information volume, you'd think that bits of information would be flowing out of every orifice (and I would be a lot smarter). But don't worry, we were issued unlimited Gigs to store all this brain data. With the aid of our trusty round watchers, us humans can distinguish over a million colors (don't try to name them). Most of us might think that a million colors are enough, but not the TV manufacturers. LCD TVs can display over 16 million colors. No doubt some tech company is working on a brain implant to let us see all 16 million colors. I'll go last.

BRAIN FAT IS BEAUTIFUL

The human brain is an amazing thingy. Like the studies of all mysteries, the more science discovers about the brain, the more questions it reveals. For years science thought we only used ten percent of our brain. Once science realized they were only using ten percent of their brains, and they invented new tests, they figured out that we use all our brain. One miniscule piece of your brain the size of a grain of sand contains one hundred thousand neurons and one billion synapses. Neurons are the brain's little social media bloggers that post messages all over your body. These little Yodas are responsible for converting thoughts into the tasks that some of us find important like talking, eating, walking, eating, thinking, eating, you get the idea. At first, science thought we were born with all our neurons. But once again, they grew a few more neurons and did more tests and they realized the brain does grow more neurons. The synapse is that mysterious connection point between two neurons, quietly passing along electrical or chemical messages like an invisible government eavesdropper. This wild network of electrical and chemical interactions occurs within a three-pound organic blob that is 40% water, protein, carbohydrates, and salts. The other 60% is brain fat (not to be confused with brain fart). This is not the kind of fat you want to lose, no matter how tight your hat fits. So, the next time someone calls you a fat head, just smile and take it as a complement. Brain fat is beautiful.

THE RING FINGER RULES

While the middle finger is always trying to get someone's attention, the ring finger might be the king finger. You may think that the decision to use the fourth finger as a ringer was a fluke because that finger just happen to be the one that fit the ring (and that ringer finger went viral). Not so. This ongoing tradition is based on ancient beliefs in Greek, Egyptian, and Hindu cultures. As the story goes, there is a vein that runs from the ring finger directly to the heart. This idea is very romantic story, and a complete lie. There is no express lane vein but, like most traditions, we just can't shake it off. But this small set back doesn't discourage a true king or queen. Some studies suggest that a longer ring finger indicates a higher exposure to male hormones in the womb (e.g the king incubator effect). Also, experts will tell you that the ratio of your ring finger length to your index finger length can indicate your athletic aptitude. This ratio is known as 2D:4D (not to be confused with R2D2). Long live the king ringer! For women, this 2D:4D ratio can bring the queen some cognitive clarity. In some studies, women with a higher 2D:4D ratio scored higher in cognitive tests in their later life years. Long live the queen ringer! So, the next time you want to send a royal message, forget the middle finger. Flip them the ring finger instead. This might just confuse them long enough for you to score an extra point or run away, whichever is applicable.

DROP AND JAW

We run, we cycle, we lift weights, we do all kinds of activities to get our muscles in shape. And we work hard at it. Crawling out of dawn's crack in the morning to work out before we go to work or, dragging our sorry butts to the gym after working our butts off all day. Some endure agonizing videos as they seek refuge in weight loss supplements. They can be heard saying, OMG, why am I listening to this self-absorbed thief for fifteen minutes as they try to convince me that I'm a fool for not buying their latest PumpYouUp berries from Africa. I could have been exercising! Some just say "screw it" and deal with the flabby consequences. But what if I told you that the strongest muscle in your body needs no exercise (and never gains weight). No doubt this is beginning to sound like a weight loss supplement video, but it's true! The strongest muscle per pound in your body arsenal is the masseter muscle, also known as your yapper. Your innocent little jaw muscle can grind your molars with 200 pounds of force. That's enough to crush even the stalest of sourdough. So, the next time someone gives you a hard time for hanging out at the gym snack bar instead of doing squats, just inform them that you are working on your masseter muscle with a vengeance! By the time they can google it, you'll be done with that doughnut and back to doing burping burpees.

PLEASE PASS THE GREAT POOP ON

How we love to eat. You can tell by the great proliferation of cooking shows now streaming on a channel near you. And if you dare click on one of those foody videos pushed to you by your social media DuJour, your cell phone will be constipated with links to cooking content. The body's process of consuming and expelling food is a wonderous marvel of muscles, fluids, and enzymes all working in concert to mine those precious minerals and nutrients and route the leftovers for some other species to consume (can I get a yuck). And as it turns out, we are loosely related to laundry detergents, as our guts use the same enzymes, like lipase and protease, to break down those stubborn brussels sprouts to get you the goods. But there have been false declarations made regarding this gastro journey. If you have fallen prey to the rumor that girls don't fart, you've been deceived by a stinky story. Everyone farts, it just a normal part of that crappy process. In fact, everyone farts twelve to twenty-five times a day (I'm likely in the twenty-five crowd). And if they're not firing SBDs (e.g. Silent But Deadly) during the day, it's happening under the cover of darkness while they are dreaming of cannon fire in the distance. So, the next time your loved one is threating to leave you if you don't stop eating beans, let them know they are equally culpable of methane distribution and their denial is just fake fart news.

SHINE YOUR LITTLE LIGHT

We are all familiar with the amazing lights of fireflies. Like little fairies in the woods, they flutter about with their glowing little butt cheek. There are other organisms that glow in the night. Some of them are Fungai beach bums surfing the ocean tides or insect river dwellers of the forests. The fancy word for this phenomenon is bioluminescence. It's just one of nature's playful little chemistry experiments. But did you know that humans are bioluminescent too? That's right. When you look at that person you're goo goo over, that glow they have might be more than infatuation. Ok, maybe it's just infatuation, as the light emitted by humans is a thousand times dimmer than our crappy eyes can detect. None the less, we are little glowers. Most of our glow comes from our forehead, neck, and cheeks (not the butt cheeks, the other ones). How do we glow? It turns out that when our cells breakdown glucose, they produce highly reactive free radicals (it's always the radical's fault). These radicals react with things like proteins, lipids, and fluorophores which eventually results in us emitting photons. We glow brightest during the day (dang, our timing is off again). No one is quite sure why we glow. Scientists say it doesn't appear to serve a purpose. But, perhaps just knowing that we glow will help shine a little more light into the world. We certainly could use it.

GROWING OUT OF CONTROL

By now we all know that we humans experience shrinkage as we get older. But a few of our body parts ignore the age finish line and grow through life's marathon to the very last day. Of course, hair is one of those culprits that requires routine attention to keep it from turning us into neanderthals at work (Note: it doesn't stop everyone). Nails are other perpetual body part that just won't stop, no matter how many times you scream from cutting them too close. Most people don't care much about these two growth addicts as they are a mild inconvenience, and they drive a thriving salon industry. Not many body parts support the glamor industry's demand side throughout the human life cycle quite like hair and nails. There are two other body parts that don't know when to quit and, typically, don't receive a warm welcome. I'm talking about the ears and nose. These round smooth and slightly fuzzy appendages just keep growing like an out-of-control mushroom patch on a forest stump. Yes, perhaps some circles admire oversized ears and noses but, they are usually confined to the elephant exhibit at the zoo. Those that can afford and stomach the precarious adventure of a facial remodel may attempt to conquer these growth bandits by force. Most of us will simply accept our ears and nose in retirement, hoping the Dumbo look comes into style and scanning the internet for a Pinocchio support group.

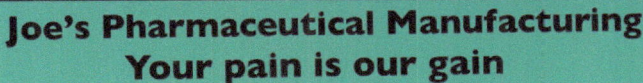

A BODY OF COLOR

Luckily, all humans are the same color on the inside. Bones, organs, blood, etc. all comply with the same color wheel. I say luckily because we seem to have enough trouble being different colors on the outside. I can't imagine the rabbit hole we'd fall in if we were different colors inside too. The reason for these colors is more of that grand body chemistry magic. Some studies suggest it might be related to the chemical composition of the typical foods we eat (which explains why those who frequent McDonalds may turn yellow and orange, like the arches). Our internal body tends to like earthy colors like brown, red, yellow, or orange. The reason our organ and tissue gangs wear their respective colors is not completely understood. Science is a flutter with trying to figure out why our liver likes brown. Maybe those colors just help us blend back into the earth when we depart like a leaf falling from the tree. That plan seems better than stumbling out of life and leaving a big purple wine stain on earth's carpet. But what about that blue blood in your veins? Won't that leave a big blue stain? Doesn't that mean I'm of wealthy royal lineage? Nice try. This little bit of trickery is where the chemist hands off to the lighting guy. Turns out that our skin is much better at absorbing red light than blue light. So, the blue light is reflected to your eyes and now you're a blue blood! That sounds like as good of an excuse as any to go shopping!

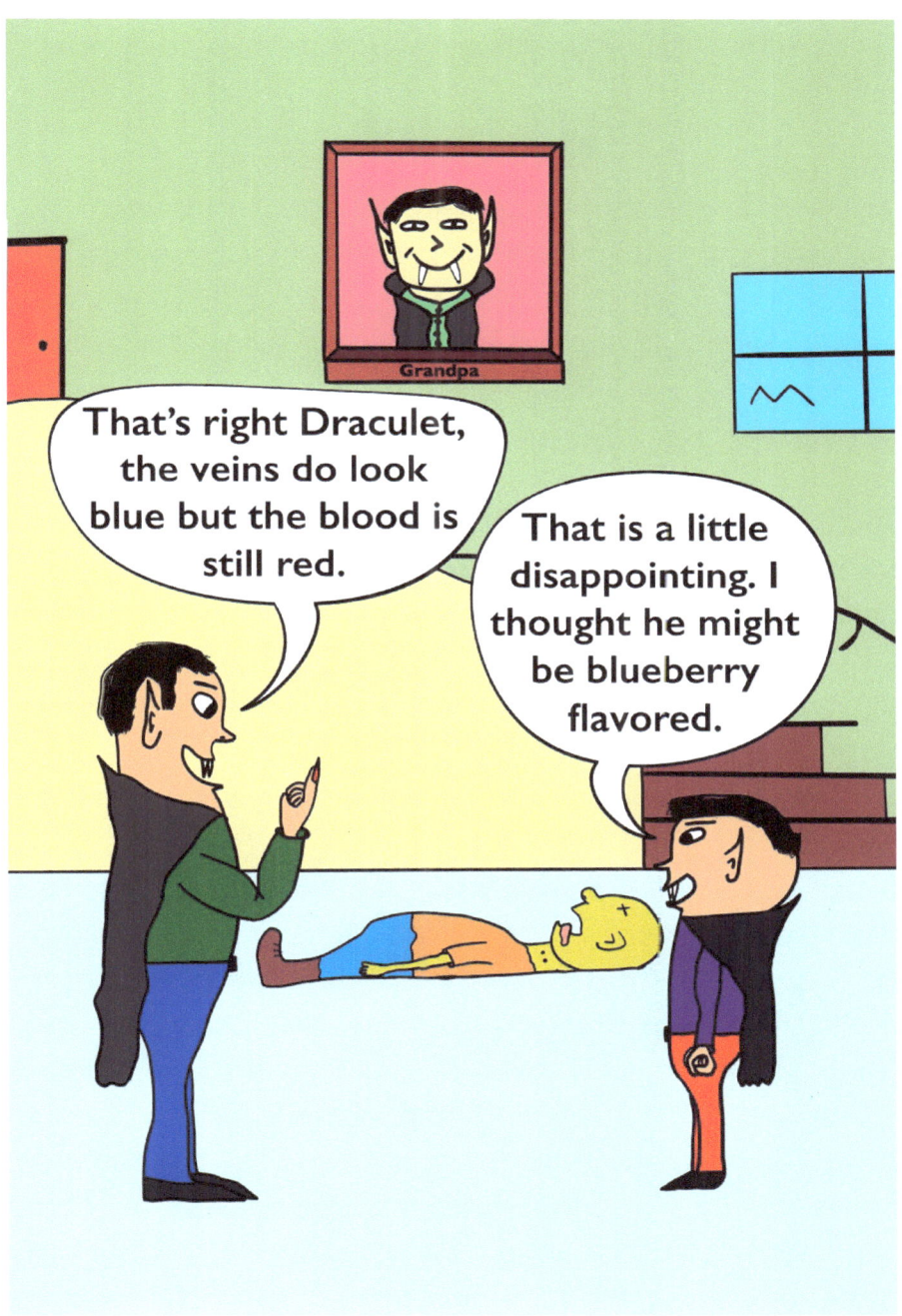

GUT SMARTS

Emotions are a funny thing, and sad thing, a lovable thing, you get the idea. Many know of the story of the neuropeptides which get released when we think and then connect with cell receptors all over our body. This is how we get that fuzzy feeling, warm feeling, feeling of anxiety, all over our body. So, all parts of the body are staying in tune and singing to the same lovely music, except our gut. Oh, our gut is singing along with the brain too but, it also has the highly complex job of dealing with all that crap we eat. As the body design decisions were being made, the gut said, "hey, if you want me to manage swallowing, do enzyme control, extract and absorb nutrients, regulate blood flow, and transport the waste out of here, then I need my own brain!" So, the body congress appropriated over one hundred million nerve cells to the GI tract to create the enteric nervous system, or ENS. This secondary nervous system is affectionally referred to as the "gut brain". It's that intelligence department within the intelligence department that no one wants to talk about, doing all the dirty jobs that keep our body republic operational. While the ENS and central nervous system do collaborate on some missions, the ENS can function independently. This independence is probably so the brain can claim plausible deniability if there is ever an investigation into that nasty business going on down there.

BELLY BUDDIES

Forget about UFOs. How about those alien bacteria visitors riding on our skin. There are approximately 1.5 trillion of those bacteria aliens to be exact. The gooier parts of our exterior can contain tens of millions of these invaders on a square centimeter of skin. Where do all these travelers come from? Well, we pet our pets, crawl around looking for contact lenses, use the same gym equipment, hug loved ones, the list goes on. But don't worry, your governing body has enlisted an army of good bacteria to fight the bad. That's right, you have the war between good and evil going on right on your nose. Before our society adopted a more frequent bathing regime, these trespassers wreaked more havoc on our health. We now manage the invasion with sophisticated weapons, like soap. We also strategically target the suspect crevasses where these squatters reside. However, one of the hidden crack houses frequently overlooked is the belly button. A recent study at Carolina State University found 2,368 different species of bacteria in the belly buttons of 60 volunteers, and every volunteer had a unique set of bacteria! Not one strand of bacteria was the same in all 60 volunteers. While that large number of bacteria may be a concern, there appears to be a bigger issue to address. There is clearly a severe lack of college belly rubbing going on at Carolina State.

MEAT HEAT

Clean energy sources are a big deal these days. Against the protest of the mole and gopher populations, people are tunneling underground to utilize geothermal energy. Wind farms are beginning to take over the windy hillsides of our country. This use of the word "Farm" seems more like a marketing ploy to give the illusion these three-armed soldiers are just part of the farm community (do they make organic turbines?). We certainly can't forget the solar panels that talk the sunshine into becoming electricity for light bulbs. This sounds like a demotion for the sun if you ask me. Perhaps we've overlooked a clean energy source that is literally right under our nose. Once again, our body's cells use chemistry and a process called thermogenesis to create heat (I feel warm already). While our organs and muscles are busy performing their jobs, the average adult is putting out 105 watts of heat per hour or 356 BTUs per hour. Our bodies produce enough heat in thirty minutes to boil a half gallon of water. And this is the energy we produce while we're resting. Imagine the energy we're generating during the New York Marathon! Who needs the electric company! With the current supply of empty office space from the work-at-home movement, we could set up tread mills in all these buildings and open free gyms. Then we could sell the energy from the body heat! This idea needs a patent. I'm on it.

UNEXPECTED RELATIONS

Many of us love the idea of "special". Those special shoes, that special outfit, that special recipe, that special talent, etc. However, when it comes to the composition of those little DNA building blocks of our physical bodies, we share a lot with a most unlikely set of cousins. It's not too surprising that we share 98% of our DNA with chimpanzees. After all, humans share many of the same features, especially as we get older. And for those that feel our cats and dogs are family, well, they are. We humans share 90% and 82% of our DNA with cats and dogs respectively. From this point, the similarities begin to get a little more disturbing. For example, you probably didn't want to know that you share 80% of your DNA with the cow that became that cheeseburger you're eating. Equally disgusting is that we share 70% of our DNA with slimy slugs. Sorry for all the drive by saltings. And not to leave vermin out of the family, we share 69% of our DNA with rats. This rat ratio is obviously higher for some people. But perhaps the most insulting fact in our molecular house is that we share more than half our DNA with chickens, fruit flies, and bananas. Chickens, fruit flies and bananas?! You've got to be kidding. How low can this go? Dear scientists, whatever you do, please don't examine our DNA similarity with fly magots. I really don't want to know. Good grief, too late.

BODY BABBLE

Oh, does the body talk, and not just with the lips. Like most creatures, we constantly use our body to communicate what we really think. For example, if you're rambling on to someone and they cross their arms over their body, this is a closed body posture, and you may have just unintentionally (or intentionally) delivered an insult. If their feet are pointing to the door instead of you, they may be planning their escape. If they maintain eye contact and tilt their head slightly, this is a sign they are engaged so, sell them something! If you are a liar and your body talk is left unchecked, the other party may see right through your rouse. For example, don't bounce your legs, tap your fingers, or start fiddling with the items on your desk like a six-year-old. Don't start changing your head position quickly like you may need an exorcism. Some studies have found people don't blink much while they are falsely telling you they didn't take that last brownie and then begin blinking faster after they have delivered the fib. So, don't start staring like store front mannequin, and then start blinking so much you look like you're having a seizure. Of course, like all communication tools, the body's directional signs are not full proof. Before you fire that liar or break off the engagement, make sure that fidget is not just their underwear riding up in their crack and the rapid blinking is just them tearing up from the pain.

THE GENE GENIE

We all know gene all too well. That gene can be a little troublemaker. Genes are the fruit hanging in your DNA that determine just how fruity you may become. They are the traits of your parents that were passed to you (without your permission) so you can make them better. At least, that's the plan. Your genes are little packages of DNA that will determine some of your physical traits and might give away the fact that you look more like that rich stud that lives next door than that guy you give the Father's Day card to each year (no one can fool gene). If you want to alter your genes, well, there is always plastic surgery. But that is just putting patches on your genes. Every cell in your body contains a copy of your genes, all 20,000 to 25,000 of them. But fear not. Your genes can change over time. It's called a mutation, a term that can bring up mental images of zombies. However, there are also good mutations. Your experiences in life and the subsequent chemicals that rain down on your genes can determine which genes are turned on and off. So, your quirky not so good behaviors could be passed to your kids, and it might take them a lot of work to find the off switch. Conversely, your rosy, uplifting, shiny attitude genes could also be passed to them, and it could take a lot of heavy metal music and horror movies to shut them off.

THE END IS NEAR

Hopefully your brain is now full of body facts, and you can go back to normal life without seeing yourself as a half bacteria half human mutant creature, too often. It should now be clear that we are all in this messy human experiment together, literally as one body, interacting with each other in unseen ways while the body's cellular generals fight the good fight. We can now tell our bosses with certainty that we'd prefer to be fired on any day but Monday and let everyone know that we now have much more respect for blood, beards, and middle fingers (hey, that's a good name for a band). We can now enjoy our new relationship with chickens and bananas, and we should stop blaming our kids for their behavior and maybe spank ourselves a little (only if you like it). Enjoy the taste of your newfound uniqueness and, going forward, be sure to include your intestinal track in all your decisions. If you are attempting to learn the language of the body, keep a dictionary handy and watch for missing periods. Remember that our physical bodies (and the rest of the planet) are not really what they look like. The technology of our eyes is a limited release (no updates available) and they can't easily reveal the truth about what we are. And if you found this book disturbing and you really do not want to commit anything to memory, for the sake of your friends and family, at least wash your belly button.

ABOUT THE AUTHOR

Ron Hine is the author and illustrator of Body Fodder and other books that tend to be about any subject that might lend itself to humoristic observation. Ron is the creator of The HAHA Train, which is dedicated to surgically extracting laughter from humans, one mind at a time. This extraction experiment is in the early stages but, we have reports of positive results on a measured number of funny bones. No mammals, animals, insects, or unknown species (that we know of) were harmed in the creation of this book.

www.ingramcontent.com/pod-product-compliance
Lightning Source LLC
Chambersburg PA
CBHW040233220526
45473CB00001B/229